Qualitative Analysis: A Guide to Best Practice

Qualitative Analysis: A Guide to Best Practice

W.A. Hardcastle
Laboratory of the Government Chemist, Teddington, UK

A catalogue record for this book is available from the British Library.

ISBN 0-85404-462-0

© LGC (Teddington) 1998

Published for the Laboratory of the Government Chemist
by The Royal Society of Chemistry,
Thomas Graham House, The Science Park, Cambridge CB4 4WF

For further information see the RSC website at www.rsc.org

Typeset by Land & Unwin (Data Sciences) Ltd, Bugbrooke, Northants NN7 3PA
Printed by Redwood Books Ltd., Trowbridge, Wiltshire

Preface

This Guide is concerned with the practice of qualitative analysis and contains advice for laboratories undertaking such work. The purpose of the Guide is to assist laboratories in establishing and maintaining a working regime which will minimise the risk of error and maximise the quality of the analytical information they produce. It is aimed primarily at laboratory managers and those responsible for any part of the analysis process from designing qualitative tests or deciding which tests to apply in particular cases to those performing the requisite experimental work.

Much of the advice here presented will be equally applicable to the field of quantitative analysis. This is inevitable given that quantitative analysis entails knowing what is being quantified.

The Guide is built upon the premise that quality begins with a thought. This means that, to ensure the end product – information – has a quality sufficient to meet a customer's requirements, thought must be given to the entire information generating process. Moreover, the mental activity of thinking must precede the practical activities of sampling and analysis.

The holistic approach adopted leads to a division of advice into three areas, *viz.*:

- sample issues;
- analysis and reporting issues;
- quality assurance issues.

The principles advanced are intended to be, as far as possible, independent of any particular technique or method of analysis.

Points are made concerning activities some would regard as peripheral to the main experimental/observational processes which produce the required data. These points are every bit as valid as those pertaining directly to experimentation. All measurements or observations are made within a particular physical environment and are recorded by a particular observer (possibly a machine). The value of a measurement, whether numeric or subjective, is therefore conditioned by the environment within which it is made and by the observer making it. Any discussion concerning analytical quality which did not address these aspects would be incomplete.

The Guide has been developed using the knowledge and experience of an advisory panel drawn from a representative cross-section of practising analytical scientists. The authors, whose contributions are gratefully acknowledged, are listed below in Table 1. This work was supported under contract with the Department of Trade and Industry as part of the National Measurement System Valid Analytical Measurement Programme.

Table 1   *Contributing authors*

Author	Affiliation
Dr R Bramley	Forensic Science Service
Dr P Brewer	Pfizer
Dr A Brown	Zeneca
Dr S Ellison	LGC
Mr W Hardcastle	LGC
Dr P Jones	Pfizer
Mr A Martin	UKAS
Dr A Strawn	Kodak

Contents

Qualitative Analysis: A Guide to Best Practice

1 This Document

This document is about principles of good practice in qualitative analysis. These principles are themselves related to one or more of the six VAM principles. In order to highlight this point, relevant VAM principles are indicated within square brackets at the beginning of each numbered section *e.g.* [VAM principle #1]. To assist the reader in locating the "Best Practice" principles, they are presented in italic with each one being further indicated by a finger sign in the left margin. Each statement of a principle is followed by some explanatory text which elaborates on the theme and may provide examples of why it is important.

The document as a whole is structured to follow the logical sequence of issues surrounding: health and safety; initial customer contact; samples; the analytical process; and quality assurance. A section on field testing is included to cover the important area of out of laboratory analyses.

2 The VAM Principles

The VAM Programme was set up with the aim of assisting analysts to obtain reliable analytical measurements [1]. To facilitate this a set of six principles has been formulated and is widely publicised. The main focus of these principles, listed below, is on quantitative measurement but it is, of course, crucially important for the analyst to know the identity of whatever it is that he or she has measured. The word 'identity' in this context has a somewhat broad meaning; it could be the name of a single substance but equally it could be a descriptor for a recognised mixture of substances *e.g.* kerosene or lavender oil.

Apart from being a necessary prerequisite to quantitation, the determination of identity is also frequently the sole purpose of an analysis. The VAM principles are equally applicable whether analysis is being performed for quantitation or for identification.

The VAM principles are:

(1) Analytical measurements should be made to satisfy an agreed requirement.
(2) Analytical measurements should be made using methods and equipment which have been tested to ensure that they are fit for their purpose.
(3) Staff making analytical measurements should be both qualified and competent to undertake the task.

(4) There should be regular independent assessment of the technical performance of a laboratory.

(5) Analytical measurements made in one location should be consistent with those made elsewhere.

(6) Organisations making analytical measurements should have well defined quality control and quality assurance procedures.

Each of the following section headings contains, where appropriate, numbered references to one or more of the above six principles. These references are intended to be indicative only, pointing to the more obvious principles which apply. Some sections could be considered to be covered by several of the principles and, in any given case, the referenced principles should not necessarily be taken to deny the relevance of any of the principles not cited.

3 Scope

The Guide has been developed as a generic document and is intended to cover a broad area of application ranging from traditional analytical chemistry/clinical chemistry to the more subjective assessments arising from activities such as organoleptic testing and document examination. For the purpose of the Guide therefore, qualitative analysis is loosely defined as follows:

Definition **Qualitative Analysis**:   The classification of objects against specified criteria to meet an agreed requirement.

In order to classify an object, in the present context, a sample of it must be made available to an analyst. The methods of classification are taken to consist of the determination of chemical or physical properties or inspection by visual or other sensory means. Some examples of classification methods are:

- Chemical tests involving colour changes, *e.g.* addition of Schiff's reagent to aliphatic aldehydes.
- Chemical tests involving production of precipitates, *e.g.* $BaSO_4$ in testing for SO_4^{2-}.
- Measurement of specific gravity.
- Observation of crystalline form or other morphological features.
- Observation of particular absorption bands in an IR spectrum.
- Limit tests in which the magnitude of a measured parameter is compared to a predetermined boundary value.

Where analysis is by reference to chemical species, all concentration levels are covered from macro to trace level.

4 Health and Safety

[VAM Principle #2, 3]

 Laboratory staff should be made aware of any potential dangers associated with the collection, analysis or storage of sample materials.

The health and safety of laboratory staff is of paramount importance and where samples are sent to a laboratory for analysis, the laboratory should ensure that it is aware of any relevant Health & Safety information connected with the sample material. This could include extracts from hazard information sheets supplied by the customer for example.

Laboratories should have a hazard assessment policy to cover all materials received; this should cover reagents as well as sample materials. All material entering a laboratory should be assigned to a predefined hazard/toxicity class upon arrival. Materials of unknown or partially known composition should automatically be assigned to high hazard/toxicity classes.

When laboratory staff collect their own samples the host organisation has a duty to ensure that such visitors are aware of any hazards connected with the environment from which the samples are collected. That is to say, from other chemicals, biological organisms or sources of electrical, kinetic, or radiative energy.

For their part, laboratory field personnel have an obligation to inform a responsible person in the host organisation if they discover, whether as a result of their tests or otherwise, that a toxic threshold has been exceeded or that a hazard or potential hazard exists in the area in which they have been working (see Section 8 on Field Testing).

5 Establishing the Requirements

5.1 An Agreed Requirement

[VAM Principle #1]

 The analyst and customer should agree on the business requirement and the technical solution.

The kind of information (chemical/physical/observational) to which this Guide relates will usually be required in order to enable the user to decide between alternative courses of action. For example:

- An industrial customer may wish to know if an impurity is present in a process feed stock so that he can avoid using it if it might poison a catalyst.
- In order to direct the course of an investigation, a police officer may wish to know if samples taken at the scene of a crime match those associated with a suspect.
- Many product specifications require that a material be present below a stated level, or between stated limits.

- It may be important for a pharmaceutical company to know that a particular chiral isomer is absent from a formulation.

These are all examples of classification, but the impact varies from economic to life-threatening. Whatever the application, however, it is crucial to ensure that the analysis performed will provide, as far as possible, the right information within the resources available and constraints operating. This can be best achieved by establishing a clear understanding between customer and analyst of the purpose of the work and of how the technical information provided will fulfil the requirement. This mutual understanding should result in an agreed statement of the requirement and the solution.

There are two important aspects to consider; the business context leading to the *customer requirement*, and the consequent *technical requirement*.

5.2 The Customer Requirement

[VAM principle #1]

 The analyst and customer need a clear mutual understanding of the requirement to ensure that the work done meets the customer's needs.

It is important for the user of analytical information to know what purpose the information is intended to serve. The analyst also should know why the customer wants the analysis performed and, in general, what use will be made of the result. Effective communication of this purpose to the analyst is important since it places the analytical problem in a context and may enable the analyst to suggest more efficient or cost-effective solutions. This communication of purpose proceeds by way of a dialogue between customer and analyst and is a vital first step in meeting the customer requirement. Factors to consider in establishing the customer requirement may include, *inter alia*:

- Economic impact of the decision.
- Effects on health or livelihood.
- Environmental consequences.
- Regulatory compliance issues.
- Criticality (for example, whether the analytical result will be the sole decision criterion).
- Timeliness.
- Cost.

Individuals in customer organisations vary in their ability to identify the information they really require. Some have considerable expertise while others, perhaps less experienced, may focus on the wrong aspects of a problem. Then again when novel situations or problems arise nobody may be too sure about what information to seek. If care is not taken at this stage then unnecessary or inappropriate analyses may be undertaken.

The first action a customer should take is to formulate a clear statement of the 'problem'. This should indicate what the business requirement is, *i.e.* what the criteria for a successful resolution of the problem would be in business terms, and should not attempt to anticipate or impose a particular technical solution. For example, the problem might be a reduced yield of a new product from a batch reactor. Analysing reactants and products for anything unusual might be a waste of time and effort if the real reason is poor heat flow and consequent unfavourable reaction conditions. The solution here relates to process control and not necessarily to feed stock composition. In this example, a business criterion might be a yield of not less than 80%. How that is achieved is a matter for the technical specialists to determine but it need not involve analysis of feed stock.

5.3 The Technical Requirement

[VAM principle #1, 2]

The methodology selected should be technically capable of satisfying the business needs.

Once the customer and analyst have a clear understanding of the business requirement, the analytical requirements appropriate to satisfying the business need can be explored and developed. To begin with, as much relevant information as possible should be obtained about the materials. In particular:

- The expected composition if known.
- The stability of the analytes of interest.
- The stability of the matrix.
- Safety information.
- The nature and likelihood of possible interferents.

A set of decision criteria (see below) will then need to be established and a suitable methodology worked out. A sampling plan may also be needed and this will be influenced by the methodology selected. In establishing the analytical requirement there are two general matters to consider; the nature of the decision criteria used, and the performance of the methods selected (including the sampling regime where appropriate).

Decisions are, in general, based on specific properties of the material or object to be classified. Properties may be physical or chemical properties, or other attributes of various kinds. Depending on the properties of interest, different kinds of criteria may be appropriate, including (but not limited to):

- *Threshold criteria*, by which a material is classified according to a measured value compared against one or more limits. Such criteria include confirmation of presence against detection limits, compliance testing against set tolerances, or simply classification thresholds chosen by prior study which, of themselves or in conjunction with other criteria, serve to identify materials. In principle, any test producing a quantitative numerical output can be reduced to a threshold criterion, including, for example, quantification of a specified component, position of a chromatographic spot or peak, measurement of physical dimensions or other properties, or calculation of match quality parameters or multivariate distance metrics.

- *Attributes*, where the material or object is classified according to the presence or absence of specified attributes, often determined by inspection. Examples include colour, shape, aroma, physical state, or morphological features.

- *Pattern matching*, in which sets of features are compared directly with reference data. Examples include direct comparison of spectra and chromatographic patterns, composition profiles, or other sets of properties.

- *Discrete classification tests*, which typically provide information of a yes/no type or, occasionally, classification into one of a few discrete categories. Examples include pH indicators, chemical 'spot tests', and many routine screening tests.

- *Interpretation*, in which observed properties or measured values are compared with properties or values predicted by established theory or from experience, typically of similar, but not identical, materials. A particularly important example of this type is structure/spectrum correlation.

Note 1: This classification of criteria is somewhat arbitrary, in that some practical criteria might reasonably be placed in different classes. For example, spectral matching might be considered a clear case of pattern matching, but if conducted using library search software providing a numerical match quality index, might be treated in practice as a threshold test. The distinction is made primarily to illustrate the range of types of evidence which should be considered, and to permit discussion of appropriate performance indicators.

Note 2: Effective classification criteria will frequently include several different types of criterion. For example, spectral matching criteria might include the specific features required to be present (attributes) and permitted ranges for their position and intensity (threshold criteria).

The selection of criteria should reflect the environment in which the tests will be applied. For example, criteria based on complex analytical tests may be inappropriate for rapid field testing. However, the criteria chosen *must satisfy the complete requirement*. It will accordingly be necessary to confirm that test methods can be practically applied as needed, or to develop practical methodologies appropriate to the situation.

Criteria should be selected to give demonstrably acceptable performance against the business needs. The performance of the methodology on the subject materials must therefore be taken into account when selecting criteria and setting decision criteria. The test method must itself satisfy performance criteria which will ensure that it is capable of establishing whether the technical criteria have been met.

It should be noted that, initially, the appropriateness of the technical criteria identified could be uncertain and in these cases a certain amount of exploratory analysis may be necessary. When this happens a potentially iterative analysis/evaluation cycle may arise with each set of analytical results feeding back to assist in redefining the technical criteria needed to achieve a business solution.

The foregoing is summarised in the Requirements and Design model illustrated at Appendix A. This emphasises the necessity of a two way communication between the laboratory and the customer and shows how the analytical requirement can be met by an appropriate technical solution consisting of a design phase and an execution phase.

6 The Sample

6.1 Collection

[VAM principle #1]

Samples should be collected in such a way as to provide a representative portion of the source material free from contamination by the sampling process. In addition, the sampling process should not contaminate the source material.

Unless agreed otherwise, responsibility for taking samples is that of the customer. The customer should be informed that, for a single sample, the validity of any inferences drawn from the analytical result will depend upon how representative the sample is of the bulk material from which it was drawn. If there are any doubts about the homogeneity of the bulk material then several samples should be taken according to a carefully worked out sampling plan. The analyst should be prepared to advise on the construction of such a plan. For each individual sample, the quantity required for an analysis is a matter for the analyst to decide; sufficient should normally be allowed for at least one replicate analysis to be made.

6.2 Containment

[VAM principle #1]

Sample containers should provide a safe and secure environment for the storage and transport of their contents. A container should not in any way alter the composition of its contents.

The purpose of a container is twofold. It serves first to maintain the integrity of its contents during transit and storage. This means preventing loss of material and also preventing contamination from the external environment. The inside of the container should be clean and free from anything that could contaminate the sample. Secondly, it protects the external environment (including people) from the sample. Care should therefore be taken to ensure that the outside of the container is free from contamination by the sample or anything else.

The sample container should be constructed of a material inert and impervious with respect to its contents. If any part of the sample container reacts chemically with its contents then, by definition, the composition of the sample has been altered and this may lead to loss (or addition) of components of interest. Note also that, apart from obvious chemical reactions, other effects such as sorption on container walls can also lead to loss of analyte.

The sample container closure should be fitted securely by the sampling personnel and, if deemed necessary, a tamper-evident seal may also be employed.

6.3 Preservation

[VAM principle #1]

 The sample should remain unaltered with respect to the unknown(s) of interest between the time of collection and the time of analysis.

The purpose of preservation is to prevent or minimise any change in the composition of a sample which would affect the unknown(s) of interest.

The stability of the sample should be considered before it is placed in the container. For example, in some cases it may be necessary to add a preservative, perhaps to prevent oxidation, whilst in others the temperature or humidity of the sample may need to be controlled. If a preservative is used care should be taken that it does not interfere with the analysis.

Note: The container itself may also have a role to play in preservation. For example, darkened glass is used to prevent/reduce photolytic decomposition and a container material impervious to air may be chosen to prevent oxidation.

6.4 Identification

[VAM principle #1]

 All samples should be uniquely identified in some way.

Since practical analysis consumes resources there is little point in analysing any sample which cannot be clearly identified and associated with a specific customer. To this end therefore staff responsible for taking samples, as well as those responsible for sample registration, should ensure that samples are clearly and adequately identified. As an absolute minimum, the following information will be required for all samples sent for analysis by a third party:

- Customer identification (who the customer is).
- Customer's unique sample ID code.

Identification will generally take the form of a label attached to the sample. In some cases identification information will be marked directly on the sample. This may occur for example when it is impracticable to attach a label.

Sample labels, where used, should be securely attached to the sample or sample container as appropriate.

All forms of identification should be resistant to attack by any hostile physical or chemical environments likely to be encountered during the sample life.

In the case of on-site analysis, discrete samples are not always taken, as for example where probes are inserted into process lines. It is nevertheless essential that sufficient information about the circumstances of a measurement be recorded so as to enable it to be used for its intended purpose.

6.5 Documentation

[VAM principle #1]

 All materials intended for analysis should be accompanied by sufficient information so as to enable their correct storage, appropriate analysis and safe handling.

Documentation consists of recorded information, which can be on any agreed medium, and which provides the analyst with information on the customer requesting the analysis, the nature of the analysis required and refers to a particular sample or samples by means of unique customer identification codes.

Every sample should be referenced in documentation giving the following information where appropriate:

- Customer identification (who the customer is).
- Customer's unique sample ID.
- Description.
- Date of sampling.
- Time of sampling.
- Location of sampling.
- Reason for taking sample.
- Nature of analysis required.
- Date result required by.
- Name of person taking sample.
- Name of person requesting analysis.
- Business address of person requesting analysis.
- Telephone/Fax number of person requesting analysis.
- Known hazards of sample material.

Documentation can consist of specially created customer or laboratory forms, faxed information, e-mail or notes of telephone conversations. The types of acceptable documentation should be agreed between the laboratory and the customer. The purpose of documentation is to provide, for each sample, a permanent record of information stating who the customer is and how they can be contacted, why the sample has been sent and identifying its point of origin both spatially and in time. This enables the laboratory to charge for the work and the customer to know what to do with the result.

6.6 Packaging

[VAM principle #1]

Where sample containers are required to be packaged for transportation, the method of packaging and the packaging materials themselves should be chosen so as to minimise the likelihood of damage to the containers or contamination of their external surfaces.

The sample container should be packaged securely so as to avoid damage during transit.

It is important to avoid damage to a sample container since not only would there be a possibility of contamination of the sample but there would also be a risk of contamination of the environment by potentially hazardous samples.

6.7 Transportation

[VAM principle #1]

The method of transportation should be chosen with regard to the stability of the sample or components of interest and any known hazards.

Responsibility for the dispatch of samples will generally be assumed by the customer. Where this is the case, the method of transport to the laboratory will be decided by the customer but the analyst should be prepared actively to advise on this where the stability of the sample is in question or a known hazard exists. In some cases the laboratory will, by arrangement, send its own personnel to collect samples.

Where a third party undertakes transport of samples, they should be fully informed of any known hazards. They should also understand the need for timely delivery if the sample or any of its components of interest can change with time.

The carrier should be aware of any special needs for control of the sample environment, *e.g.* temperature control, and ensure that these are met.

6.8 Receipt

[VAM principle #1,6]

Laboratories should have a clear and workable procedure for the reception of samples.

Sample packages should be opened as soon after receipt as possible and their contents checked. Damaged or leaking containers should be noted and appropriate local procedures followed or advice sought.

Sample registration staff should ensure that a unique identification code is associated with each and every sample. They should then check that each sample matches a documented description. If a laboratory's own identification codes are to be associated with samples, these should be attached at this stage. Where multiple containers of the same sample are submitted, the laboratory identification codes should nevertheless distinguish between them.

Poorly labelled samples or those whose origin or purpose cannot be readily ascertained should be put to one side and steps taken to resolve the difficulties.

6.9 Opening

[VAM principle #6]

Sample containers should be opened in a manner which maintains the integrity of their contents and ensures the safety of staff.

In order to minimise the possibility of contamination, sample containers should normally only be opened when the analyst intends to start work on the contents.

Care should be taken to ensure that the lids/caps/closures of sample containers are not placed contents side down on surfaces that may contaminate them.

Care should be taken when opening sample containers that may be under pressure. This is both from a safety point of view and to minimise the risk of contaminating equipment that may be used for other analyses.

Care should be taken when opening sample containers that may be under partial vacuum. This is both from a safety point of view and to minimise the risk of contaminating the sample.

For some materials, sample containers may need to be maintained at temperatures above or below ambient. Here too, caution should be exercised when opening such containers.

The sample container should be opened only long enough to extract the required amount of sample. To minimise the risk of contamination the closure should then be replaced immediately and securely.

Where it is known or anticipated that a sample contains hazardous material, appropriate precautions should be taken for its management.

6.10 Sub-sampling

[VAM principle #6]

Where it is necessary to take a sub-sample from an existing sample, such sub-samples should be uniquely identified and should be traceable to the parent sample.

Samples sometimes need to be split when they are to be analysed for more than one constituent. For example, different parts of a laboratory may specialise in different analytical techniques and may require their own portion of the sample to work on.

Sub-sampling should, in general, be carried out so as to obtain as representative a portion of the parent sample as possible and without contaminating either the parent sample or the sub-sample. For multi-phase samples the purpose of sub-sampling may be to obtain a portion of a single phase in which case the sub-sample should be representative of the phase of interest. Any hazard information relating to a sample should be considered when sub-sampling.

All analytical results obtained on sub-samples should be clearly linked to the parent samples.

6.11 Storage

[VAM principle #6]

Material received for analysis should be stored in a safe and secure manner and under conditions which preserve its integrity.

Samples should be stored under secure stable conditions which do not alter the composition of the samples. This advice applies from the moment a sample is taken and for the lifetime of the sample.

Some samples may degrade naturally during storage and therefore will have a limited useful lifetime.

Where a customer is aware of any special storage requirements for a sampled material, this information should be documented and communicated to the laboratory. For its part, and in the absence of any specific storage information from the customer, the laboratory should satisfy itself that its normal storage regime will be appropriate. This may entail contacting the customer for clarification.

6.12 Disposal

[VAM principle #1, 6]

Laboratories should have a policy covering the retention and disposal of analysed samples.

Few laboratories have the facility to retain analysed samples indefinitely. Each laboratory should have a policy for disposal of sample remnants once an analysis has been completed and the results reported to the customer. This should include the length of time the sample will be retained for and the method of its disposal.

Some customers may wish to have analysed samples returned especially if they are of high value and non-destructive methods of analysis have been employed. Alternatively there may be other commercial or legal reasons for returning sample remnants.

Where the customer has no further interest in his samples then disposal will normally be the responsibility of the laboratory. The laboratory should satisfy itself that it has all pertinent safety information relating to a sample and that its disposal procedures comply with any legislation covering such activities.

Apart from any agreement with the customer, a laboratory may be required to retain samples for a minimum period stipulated by an accredited QA protocol.

7   The Analytical Process

7.1 Classification Criteria

[VAM principle #1]

The classification criteria should be sufficiently well defined to enable an unambiguous result to be obtained by appropriate methods.

Qualitative analysis is based on *criteria*. Criteria may be, for example, the presence of a particular analyte above a specified level, physical properties within particular limits, a match between two spectra, or combinations of features. Whatever the nature of the criteria, however, it is important that the criteria be unambiguous, clearly stated and, as far as is possible, objective. For example, 'the melting point should match the reference value' is insufficient because it does not specify the degree of match acceptable. 'The melting point should match the reference value *within 1 °C and the material should melt entirely within a range less than 0.5 °C*' is a clear, unambiguous and objective statement. Acceptable criteria will depend on the nature of the criteria in use. For example, using the terminology at Section 5.3:

- *Threshold criteria* normally take the form of stated numerical thresholds or ranges against which a material is classified. The limiting values chosen and the specification of the criteria should take appropriate account of uncertainties in determined values. In particular, the criteria should specify how uncertainty information should be considered in interpretation (including the case where classification is made on the result as obtained).
- *Attribute* criteria will generally consist of clear descriptions of the relevant attributes or reference artefacts (for example, colour reference panels) which demonstrably permit unambiguous classification by comparison.
- *Pattern matching* criteria will normally indicate the relevant reference data, the basis of the comparison (such as visual inspection, the set of features for consideration, or computational comparison methods) and the required quality of the match. The latter might be conveyed through numerical criteria or by example (such as a diagram showing acceptable and poor matching).
- *Discrete classification* criteria will typically describe or provide examples of the expected behaviour of the test for particular cases.
- *Interpretation* criteria will typically specify the basis for prediction and the quality of match between prediction and observation. It may additionally be necessary to specify appropriate reference data and, critically where expert judgement is required, the level of training and experience of the analyst.

Whatever their nature, the criteria should demonstrably distinguish adequately between different classes and permit unambiguous assignment of an object or material.

The criteria used to interpret each category of observational or experimental data should be recorded and available for reference by relevant staff as required.

7.2 Method Selection

[VAM principle #2]
All methods used for qualitative analysis (and also for quantitative analysis) should be documented and fit for their intended purpose.

Methods should be chosen which enable the established test criteria to be evaluated. In other words the kind of information they provide should clearly be of use in meeting the customer requirements.

The scope of a method should define *inter alia*:

- The analytes that respond under the conditions of the test.
- The features of the analyte response which enable it to be identified.
- The lowest concentration above which the analyte can be reliably identified.
- Known interferents.
- Known matrix effects.

It is important that the latest version of a method be used.

7.3 Method Validation

7.3.1 Specificity

[VAM principle #1, 2]

 *The specificity of a test for identity should always be known.*

Tests for identity vary in their ability to discriminate between similar materials. The specificity of all such tests should therefore be established before use by means of validation studies.

Tests with high specificity are, in general, preferable to those having lower specificity since there will be less likelihood of misidentification. Tests should be chosen however with due regard to the end use of the information generated. For screening purposes, for example, a test with a relatively low specificity may well be perfectly acceptable.

7.3.2 Detection Limit

[VAM principle #1, 2, 6]

 Any qualitative test employed must be sensitive enough to detect the species of interest at the concentration levels of interest.

Studies should be undertaken to show that the species of interest is reliably detected at the level of interest. Typically, such studies examine the lowest level at which a test reliably indicates the presence of a particular species; that level is generally referred to as the detection limit.

7.3.3 Misclassification Rate

 [VAM principle #1, 2, 6]
Rates of misclassification must be known and controlled.

It is particularly important, in the context of qualitative analysis, that the analyst has reliable information on the rates of both false positive and false negative responses associated with the tests employed. A procedure should be in place to ensure that such information is obtained and available for all tests in use.

7.4 Computations

[VAM principle #3, 6]

 There should be a policy for reviewing computational procedures and checking the correctness of results obtained.

Calculations are frequently needed even for qualitative work; for example, reagents have to be made up to particular concentrations or threshold response levels calculated. It is good practice to have an independent check made on all calculations.

Where computers are used to perform calculations, the underlying computational procedures should be known at least in outline. It should be evident that computations using these procedures will in principle produce the desired output. If possible, evidence of correct operation of the algorithms employed should be obtained by feeding in a range of test data and comparing the results with those obtained by independent calculation. Several cases are known where the wrong formula has been encoded in a computer program and so complete reliance should never be placed on the correctness of any piece of software however well known or prestigious its manufacturer may be.

Special care should be taken when computer assisted interpretations are undertaken. The use of computerised libraries and databases can be problematical if the analyte of interest is not contained within them, and in any case the 'best match' is not necessarily the correct classification. Additional criteria may accordingly be necessary. Similarly, expert systems do not provide proof of identity; they indicate likelihood on the basis of a particular training data set, and are subject to the same limitations outlined above. It is important to ensure that their scope of application is not exceeded.

7.5 Confirmation

[VAM principle #2, 6]

If appropriate, a confirmatory test should be employed to substantiate the conclusions from the primary test or tests.

It is important that any confirmatory test used is completely independent of the primary test. It should also be noted that, just as it is possible to obtain a false positive result, it is also possible to obtain a false negative result. The use of confirmatory tests is therefore not restricted to confirming only positive results.

Where interpretations are based upon more subjective observations (such as handwriting analysis or microscopic observations), and depending upon the use to which the results will be put, it may be necessary to have a second competent person repeat the observations and verify the original conclusion(s).

8   Quality Assurance

8.1 Environment

[VAM principle #6]

 The working environment should be actively controlled to ensure that all relevant parameters are within appropriate limits.

The working environment should be conducive to good analytical science. Things to consider are:

- A clean and tidy working area.
- Clean and tidy surroundings.
- Adequate ventilation.
- Adequate illumination.
- Stable and controllable temperature.
- Stable and controllable humidity.

These are factors which can directly or indirectly affect the quality of results and can be controlled by good housekeeping and good building management.

An anti-contamination protocol should be established and adhered to and consideration should be given to the need for special clothing and/or the reservation of equipment for particular purposes.

If spillage of analytical material occurs during the course of an analysis, either through accident or breakage, it must be immediately removed and the affected area decontaminated. An appropriate strategy should be in place for confirming the effectiveness of the decontamination procedure. It is recommended that a log of such incidents, however minor, be maintained.

Where an organisation routinely undertakes analysis for a given analyte at both macro and trace levels, separate, and separated, facilities must be provided for each type of work. The movement of staff and equipment between these areas should be strictly controlled.

For trace level work in particular, it is recommended that all glassware used should be reserved for that purpose only. In addition, the local environment should be monitored for the analyte of interest.

8.2 Equipment

[VAM principle #2]

 Equipment used in the course of an analysis should be well maintained and operated within its design parameters.

All containers and transfer vessels employed during an analysis must be inspected before use to ensure that they are free of obvious contamination. It is good practice to clean such apparatus thoroughly as soon as possible after use. Cracked or chipped glassware should on no account be used; it is a source of potential contamination but more importantly it is, quite simply, dangerous.

Electrically powered instruments should be allowed sufficient time to warm up and stabilise if not kept permanently switched on.

All equipment employed in analysis should be regularly serviced and maintained. A log should be kept for each instrument showing details of any permanent adjustments made; this should include who made the adjustments, when and for what reason.

In addition to being maintained, equipment may also need to be calibrated and here traceable reference standards should be used where available. A separate log should be kept for the purpose and should show the date of the calibration together with all relevant data. A calibration frequency should be determined for such equipment and this should be stated in the front of the log. Equipment subject to calibration should bear an easily visible label showing the calibration status.

Equipment parameters should be checked before use to ensure they are set as stated in the method.

All equipment should be subjected to 'fitness for use' tests. For equipment in regular use the tests can be carried out at pre-set intervals. Where equipment is employed sporadically however such tests should be applied before each use.

8.3 Reagents

[VAM principle #2, 6]

 The purity of all reagents employed in a method should be known, at least approximately, and all materials used should be tested to ensure they do not contain substances which would interfere in the analytical method.

A blank test should be performed on all new batches of reagents (including solvents and carrier gasses) to guard against the possibility of a false positive result being produced by an impurity. Reagent blanks should always be included with each batch of samples analysed; this will reveal any contamination or degradation of the reagent over time.

The shelf life of all purchased reagents should be known and expiry dates checked before use. In addition, made up reagents should be labelled with the date of preparation and an expiry date. Other appropriate information, especially for GLP purposes, could include the name of the person who prepared the reagent and a lot number. Labels should also show what solvent the reagent is in and its concentration.

Just as with samples, reagents too can be affected by the conditions under which they are stored. Care should be taken therefore to ensure that all reagents are stored under appropriate conditions.

The quality of a purified water supply should not be assumed to be adequate. Instead, appropriate tests should be carried out to verify that it *is* adequate. The equipment producing such water may well yield a good quality product but the purity can be compromised by the storage or transmission systems. For example, plasticisers can leach from pipework and cause problems in trace analysis work [4].

Purity is not the only parameter affecting the quality of a solvent. As an example, for some purposes, the boiling range of a solvent may be important as well. Method documentation should therefore, where appropriate, contain a section specifying the required solvent properties.

8.4 Operators

[VAM principle #3]

 All analyses should be conducted by personnel who have demonstrated the relevant competencies.

Training records should be maintained for all staff contributing towards analytical reports and evidence should be available of their continuing competence.

8.5 Documentation of Methods

[VAM principle #2]

 Analytical methods should be documented sufficiently well as to enable their successful use by competent analysts unfamiliar with them.

The documentation for all qualitative methods should be carefully reviewed to ensure that they are internally consistent and free from ambiguity. This includes any guidance notes on interpretation.

All methods should, in any case, come under a regular review cycle to ensure that, for example:

- They are not drifting out of scope due, for example, to changing customer requirements.
- They remain compliant with legislation, *e.g.* do not use solvents restricted or banned under the Montreal Protocol.
- They remain economic to perform.

A procedure for performing and documenting method reviews should be established.

8.6 Laboratory Records

[VAM principle #2]

 A record should be kept of all observations made and of all data generated during an analysis. All information necessary to perform the analysis should also be recorded.

All data produced during the course of an analysis and any other relevant information should be recorded. Data may consist of the digital output of instruments or manual observations such as analogue readings, colour changes or state changes. Information about instrument parameter settings, date/time and personnel is also pertinent to every analysis and should be noted.

Deviations from standard procedures should be noted and explained and any observations of unusual events should also be logged.

Adherence to this principle should allow any competent analyst familiar with the analytical method to reproduce the results of an analysis performed by someone else. It will also provide an audit trail so that, should a query arise concerning a result, the necessary facts will be available and the steps leading to the result can be reconstructed on paper.

Laboratories should have a policy covering what information will go into a record and the length of time for which records will be kept.

8.7 Quality Control

[VAM principle #6]

Quality control measures should be in place which cover all activities likely to affect a result.

All methods employed for qualitative (or quantitative) analysis should be operated under a quality control regime. Regular check samples should be analysed to ensure that the expected experimental responses and corresponding interpretations are obtained. Check samples should include blanks and reference standards.

It is good practice to have one or two check samples containing the analyte of interest at concentrations just above and just below the limit of detection. These should be put through the analytical procedure at regular intervals and certainly when new batches of reagents are prepared, to confirm method performance.

8.8 Independent Assessment

[VAM principle #4]
There should be independent assessment of performance.

One way of obtaining an independent assessment is to participate in Proficiency Testing (PT) schemes. PT schemes permit laboratories to compare their analytical performance with that of other laboratories and, as such, are a valuable tool for maintaining or improving the quality of results.

Participation in PT schemes (where they exist) covering the qualitative tests in use is always beneficial, especially for tests involving largely subjective components *e.g.* smell, shape recognition *etc.*

To be of any real value, it is important that PT scheme samples are not treated any differently from normal samples. Also, any deficiencies exposed by scheme reports should be investigated and remedial action taken. This action should always include consideration of whether the deficiency could have affected results already reported.

Another way in which laboratories can establish independent assessment of their performance is to acquire accreditation to a third party quality system such as GLP, systems based on ISO Guide 25 or ISO 9001. Again, it is important that any (agreed) deficiencies noted by the independent auditors are rectified.

9 Interpretation of Results

[VAM principle #3]

 The interpretation of observations or experimental data should provide a classification consistent with the available information.

Interpretation of observations or experimental data also requires some skill and in some instances, considerable skill. If reliable observational or experimental data have been obtained they still have to be interpreted in the context of the analytical problem and this may need skills different to those required for obtaining the data. If such techniques are employed then appropriate evaluative skills must be available to minimise the risk of misinterpretation of the data.

Interpretation should be consistent with the observations, the established criteria (Section 7.1) and relevant quality control and assurance data. Any variation from the established criteria should be recorded and the reasons for the variation noted. It is important that the thought processes that lead to reported conclusions are documented.

10 Analysis Report

[VAM principle #1]

 The analysis report should clearly indicate to the customer how the analysis has met his/her requirements.

The test report issued to the customer should identify:

- The laboratory.
- The customer.
- The sample.

In addition,

- The result of the analysis should be clearly stated and linked to the customer's sample identification code or to the laboratory's unique sample reference code.
- The date of the report should be shown.
- The signature of the person responsible for the report should appear on the report.
- The name of the person responsible for the report should be PRINTED LEGIBLY next to the signature.
- The report should include all relevant information as well as an indication of the reliability of the conclusions and any caveats thought advisable.

Extra items of information can appear on the report either by agreement with the customer or at the discretion of the laboratory. Examples of such information can be found in Guide 25 of the International Organisation for Standardisation [2]. See also the EURACHEM Guidance Document No 1 [3].

11 Field Testing

[VAM Principle #2, 5]

☞ *Field testing is subject to additional problems to those found in the laboratory and staff working in the field should be aware of these before they set out.*

If analyses are conducted in the field then consideration must be given to the special problems which can arise in these situations. For example, control of the environment may be a problem – most commonly through the need to perform tests within a narrow temperature range.

Transporting equipment can affect its calibration. Where qualitative analysis in the field involves quantitative measurements, special attention should be paid to the calibration of any equipment used.

Of particular concern in field testing is the possibility that materials employed in testing may contaminate either the bulk material which is being tested or the local environment. Staff engaged in field testing should ensure that they take all reasonable steps to prevent such contamination occurring.

12 Further Information

Further information underlining the advice presented in this Guide can be found in the publications listed in the Bibliography. *The Manager's Guide to VAM* is intended to assist laboratory managers implement the VAM principles. ISO Guide 25 is an international quality standard and sets out the procedures laboratories must follow in order to acquire and maintain accreditation to this standard (administered by UKAS in the UK). The EURACHEM Guidance Document supplements ISO Guide 25 and provides detailed help on interpreting the latter. The publication *Achieving Quality in Trace Analysis* covers the special problems connected with this particular area.

The *VAM Bulletin*, published quarterly, is the official organ for disseminating VAM related information which includes articles contributed by readers. Additional information on these publications or anything else to do with the VAM Programme can be obtained from the LGC VAM Helpdesk on +44 (0)181 943 7393.

Appendix A: Requirements and Design Model

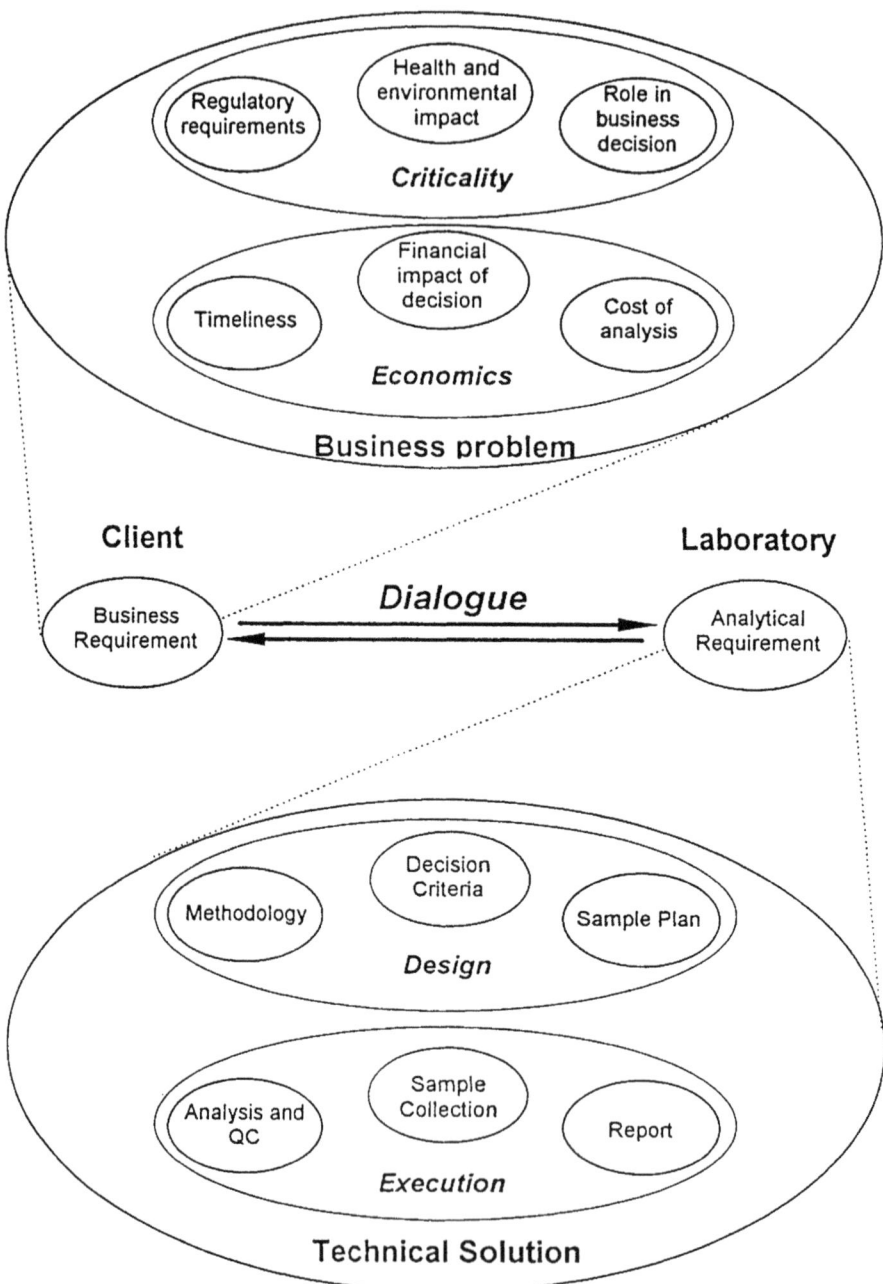

Appendix B: Bibliography

1. The Manager's Guide to VAM, Issue 1, 1996.
2. International Organisation for Standardisation, Guide 25, 3rd Edn., 1990.
3. EURACHEM Guidance Document No. 1, 'Accreditation for Chemical Laboratories', 1993.
4. Guidelines for Achieving Quality in Trace Analysis, The Royal Society of Chemistry, Cambridge, 1995, ISBN 0-85404-402-7.

www.ingramcontent.com/pod-product-compliance
Ingram Content Group UK Ltd.
Pitfield, Milton Keynes, MK11 3LW, UK
UKHW052231190226
468218UK00014B/36

9 780854 044627